DID YOU KNOW?
ANIMAL ADAPTATIONS

I0812957

20 THINGS YOU DIDN'T KNOW ABOUT

MAMMAL ADAPTATIONS

SLOANE HUGHES

PowerKiDS press
New York

Published in 2023 by The Rosen Publishing Group, Inc.
29 East 21st Street, New York, NY 10010

Portions of this work were originally authored by Kristen Rajczak and published as *20 Fun Facts About Mammal Adaptations*. All new material in this edition was authored by Sloane Hughes.

Editor: Amanda Vink
Book Design: Tanya Dellaccio

Photo Credits: Cover (speech bubble) Gohsantosa/Shutterstock.com; cover (logo art) NazArt/Shutterstock.com; cover isabel kendzior/Shutterstock.com; p. 5 (meerkats) Mrinal Pal/Shutterstock.com; p. 5 (mountain goat) Karla Ferro/Shutterstock.com; p. 6 Alexey Seafarer/Shutterstock.com; p. 7 Andrew Sutton/Shutterstock.com; p. 9 (hair) Designincolor/Shutterstock.com; p. 9 (dog) Olena Yakobchuk/Shutterstock.com; p. 10 McLey/Shutterstock.com; p. 11 Jeannette Katzir Photog/Shutterstock.com; p. 12 BlueBarronPhoto/Shutterstock.com; p. 13 Krasula/Shutterstock.com; p. 14 Paul Loewen/Shutterstock.com; p. 15 Danita Delimont/Shutterstock.com; p. 16 FloridaStock/Shutterstock.com; p. 17 Helen E. Grose/Shutterstock.com; p. 18 Tom Reichner/Shutterstock.com; p. 19 sculler/Shutterstock.com; p. 20 (mouse) MelaniWright/Shutterstock.com; p 20 (den) Gerald Peplow/Shutterstock.com; p. 21 LouisLotterPhotography/Shutterstock.com; p. 22 teekayu/Shutterstock.com; p. 23 Lisa Hagan/Shutterstock.com; p. 24 Lukas_Vejrik/Shutterstock.com; p. 25 Debbie Steinhausser/Shutterstock.com; p. 26 nigel baker photography/Shutterstock.com; p. 27 (kangaroo) Michal Pesata/Shutterstock.com; p. 27 (echidna) Ken Griffiths/Shutterstock.com; p. 29 Thampitakkull Jakkree/Shutterstock.com.

Library of Congress Cataloging-in-Publication Data
Names: Hughes, Sloane, author.
Title: 20 things you should know about mammal adaptations / Sloane Hughes.
Other titles: Twenty things you didn't know about mammal adaptations
Description: New York : PowerKids Press, [2023] | Series: Did you know? animal adaptations | Includes index.
Identifiers: LCCN 2021055892 (print) | LCCN 2021055893 (ebook) | ISBN 9781538386835 (library binding) | ISBN 9781538386811 (paperback) | ISBN 9781538386828 | ISBN 9781538386842 (ebook)
Subjects: LCSH: Mammals–Adaptation–Juvenile literature. | Adaptation (Biology)–Juvenile literature.
Classification: LCC QL706.2 .H85 2023 (print) | LCC QL706.2 (ebook) | DDC 599–dc23/eng/20211118
LC record available at https://lccn.loc.gov/2021055892
LC ebook record available at https://lccn.loc.gov/2021055893

Manufactured in the United States of America

CPSIA Compliance Information: Batch #CSPK23. For Further Information contact Rosen Publishing, New York, New York at 1-800-237-9932.

CONTENTS

MIRACULOUS MAMMALS

All mammals are **vertebrates** and have some form of hair or fur. Most mammals give birth to live young and feed them with milk from their body. There are more than 6,000 different species, or kinds, of mammals!

Mammals live around the world in just about every **environment** you can imagine. Over a long period of time, animal bodies and actions tend to change to make each species better suited to its habitat. These changes are called adaptations.

Some mammals call bitter cold areas their home. Some mammals prefer to live where it's hot all the time!

THE HEART OF THE MATTER

DID YOU KNOW?

Mammals have special hearts that help them maintain body temperature.

Mammals have a heart with four chambers, or parts. The four chambers have a special purpose. They keep oxygen-rich blood and oxygen-poor blood separate as the lungs do their work. Because this blood doesn't mix, it's easier for a mammal to maintain its body temperature.

Blue whale babies, called calves, can drink roughly 50 gallons (189 L) of milk a day.

DID YOU KNOW?

Blue whales are most likely the largest animals that have *ever* lived on Earth.

Blue whales can reach 110 feet (34 m) long! Even though blue whales live under the water, they're mammals. They have backbones, they have a little hair, they breathe with lungs, and they feed their young milk.

A HAIRY BUSINESS

DID YOU KNOW?

Hair is tough stuff!

Both hair and fur are made of the same thing: keratin. This is a type of **protein** that also makes fingernails. In addition, keratin makes up animal hooves, horns, feathers, wool, and some skin cells. Even though fur and hair can be soft, things made of keratin are some of the toughest nonminerals found in nature. Keratin can flex, or bend, without tearing.

TYPES OF HAIR GROWTH IN MAMMALS

Tough outer-layer guard hairs stay in the anagen stage for a long time, so they need trimming!

ANAGEN

TIME WHEN A NEW HAIR BEGINS TO GROW

CATAGEN

TIME WHEN A HAIR STOPS GROWING

TELOGEN

TIME WHEN A HAIR IS SHED AND A NEW HAIR FORMS IN THE **FOLLICLE**

DID YOU KNOW?

Only mammals have hair, but it comes in many types!

Did you know that a cat's whiskers are a type of hair? Cats use their **sensitive** whiskers to help them measure distance and **navigate** in the dark. Walruses use their thick whiskers to find food. Dogs and other animals use their whiskers to show their mood.

Each mammal hair grows from a hair follicle.

The scientific name for this reaction is piloerection, but we know it better as goose bumps!

DID YOU KNOW?

Hair can help a mammal stay warm in more ways than you think!

When some mammals, such as chimpanzees, are cold, their hair raises and stands on end. This traps air between the layers of skin and hair, helping to keep the animal warm. This also happens when the animal feels **threatened**, but for a different reason. It makes the animal look bigger than it is!

DID YOU KNOW?

Sea otters have the thickest fur of any animal.

Sea otters have between 850,000 and 1 million hairs on each square inch of their body! Sea otters swim in the Pacific Ocean, where temperatures can get really cold. They have two layers of fur that help to keep them warm. The outer layer is thick and stiff, and the underlayer is soft.

Because sea otter hair is so **dense**, it's water **resistant**!

DID YOU KNOW?

Some animals' fur changes color depending on the season.

Stoats, or short-tailed weasels, change color in the winter. A stoat's fur is normally a dark brown color, but it changes to white to blend in with the snow. That makes it much harder for predators to figure out their location.

OUT IN THE COLD

DID YOU KNOW?

Arctic hares have small ears to help them stay warm.

Arctic hares adapted to have small ears because shorter ears mean there is less body surface to keep warm. Arctic hares also have thick fur, small noses, and a lot of body fat. Just like stoats' fur, their fur changes to white as winter approaches.

Arctic hares are typically shy, but when it's really cold they sometimes gather together to keep warm.

On the coldest winter days, deer mice go into a state called torpor. This is a state of inactivity in which their body temperature drops to 68°Fahrenheit (20°Celsius). The mice group together very closely to share body heat.

DID YOU KNOW?

Deer mice can lower their body temperature in winter to conserve energy.

DID YOU KNOW?

Polar bears have clear fur and black skin.

Polar bears are well-known as big, white mammals. But did you know that the hairs of their fur are actually clear? Also, their skin is black under all that fur. That helps them **absorb** as much heat from the sun as possible. Each hair of their fur is also hollow, or empty, which helps to keep heat close to the bear's body.

Even though it's really cold where they live, polar bears often overheat. To cool off, they roll around in the snow.

DID YOU KNOW?

Some animals, like the Eastern gray squirrel, hide away lots of food to eat during the winter.

In the fall, squirrels busily collect nuts, seeds, and other foods. They hide these things in many different places called caches. Sometimes they pretend to hide things in a certain place just to throw other squirrels off! When winter comes, they remember where to go to find their hidden snacks.

DESERT LIVING

DID YOU KNOW?
Many desert animals have an active nightlife.

When living in the dry desert heat, it makes sense for lots of mammals to be most active at night. This is called being nocturnal. Some mammals, like coyotes, are crepuscular, which means they are most active at dawn and dusk. Their prey is also active during that time of the day.

AMERICAN BADGER

When a camel does find water, it can drink a lot! Some sources say a camel can drink more than 32 gallons (121 L) of water at one time.

DID YOU KNOW?

Even in hot desert temperatures, a camel can survive for nearly a week without water.

Camels have humps, but these humps are not used for storing water. Instead, they store fat. A camel can go for a long time—sometimes several months—without eating. The fat in their humps gives them energy when they can't eat or drink. Camels store more water in their blood because of special cells.

DID YOU KNOW?

Pack rats create strong nests out of desert **debris** and their own urine!

Pack rats build structures called middens. These can be up to 3 feet (0.9 m) high and 8 feet (2.4 m) long. They're made from plants, sticks, stones, shells, and bones. The rats use their urine, or pee, to create a natural cement to harden the midden. Yuck!

Pack rat middens are hard as rock, and they can last thousands of years.

DID YOU KNOW?

The Cape ground squirrel carries its shade wherever it goes!

The Cape ground squirrel is native to South Africa and lives in small groups. These squirrels burrow, or dig, in the dirt to make homes. When the sun gets hot, the Cape ground squirrel uses a unique adaptation. It fluffs up its big, bushy tail and uses it to block the sun's rays.

MAMMAL DEFENSE

DID YOU KNOW?

North American porcupines have 30,000 or more quills!

Quills are large, hollow, stiff hairs. Porcupines try to scare predators by shaking these quills, which causes them to rattle against one another. If that doesn't work, a porcupine will turn its back to the animal bothering it and swing its tail. Those quills can cause a lot of harm!

Opossums don't play dead on purpose. It's an **involuntary** reaction to danger. It can take a few minutes or up to a few hours for opossums to become active again.

DID YOU KNOW?

Opossums "play dead"!

Opossums will fight back if they are threatened. But if that doesn't work, they fall to the ground and stay very still. In fact, they can't move. They don't blink, and they don't respond to pain. This is an adaptation to help an opossum survive because many wild animals will not eat an animal that appears to be dead.

DID YOU KNOW?

Some mammals use **venom** to protect themselves against predators.

Male platypuses have spurs, or pointed parts, on their back legs that are venomous. Though the venom is not lethal to humans, it's very painful. Other venomous mammals are the European mole and the Northern short-tailed shew.

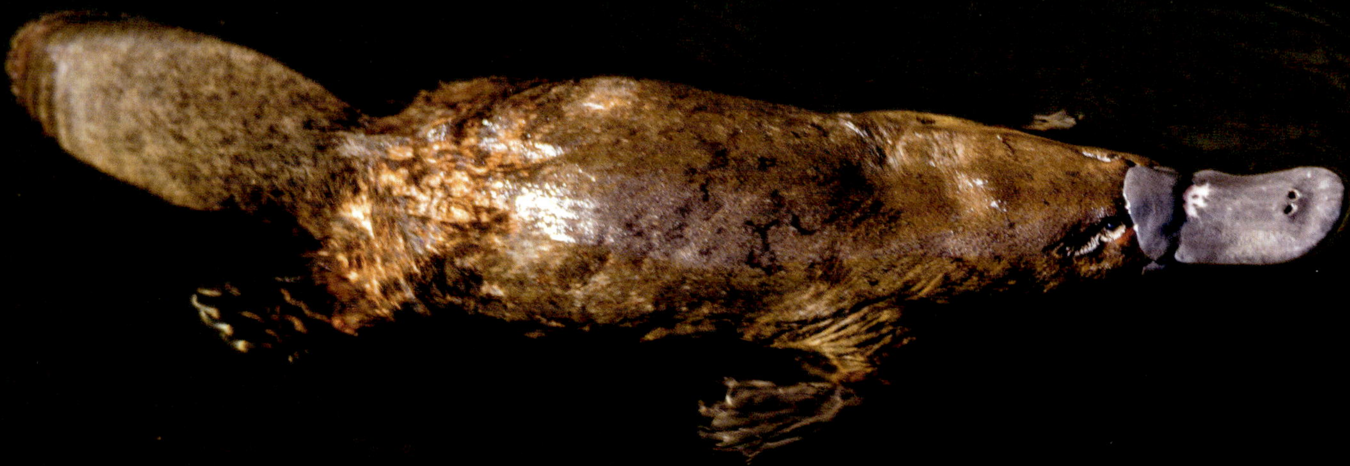

DID YOU KNOW?

A skunk's foul-smelling spray can cause blindness for a time.

Skunks, weasels, and ferrets all spray a liquid called musk to keep predators away. It's stinky! Skunk musk is the most powerful. It's so strong that it can cause temporary blindness in any animal unlucky enough to get sprayed.

DID YOU KNOW?

Only two mammals lay eggs!

Platypuses and echidnas are both monotremes, or egg-laying mammals. But all mammals—from the small bumblebee bat to the large blue whale—feed their babies by producing milk.

MAMMALS HAVE THEIR BABIES IN ONE OF THREE WAYS.

PLACENTAL MAMMALS

- baby grows inside the mother's body
- baby is connected to its mother through an organ called a **placenta**

MONOTREMES

- lay soft-shelled eggs
- young drink milk from pores on mother's body

MARSUPIALS

- babies are very tiny when they are born
- babies develop in a mother's pouch

A FAMILIAR MAMMAL!

You're very familiar with one type of mammal—humans! Like other mammals, people have developed adaptations depending on where they live. For example, people who live on the Tibetan plateau in Asia have adapted to living at higher altitudes, or height above sea level. These people breathe faster than those who live at sea level.

Adaptations help mammal species survive over time. These adaptations allow mammals to thrive all across the globe. What other mammal adaptations can you think of?

The family members in this photograph are wearing traditional Tibetan clothing.

dense: Having parts that are close together.

environment: The conditions that surround a living thing and affect the way it lives.

follicle: A tiny hole in the skin from which hair grows.

involuntary: Not done by choice.

navigate: To find one's way.

placenta: The organ in most mammals by which the unborn baby is joined to the mother in order to receive nourishment.

protein: A basic structure made of amino acids and found in all life. There are many different kinds.

resistant: Not affected by something.

sensitive: Able to sense very small changes in something.

threatened: Likely to be harmed.

venom: A type of poison made by an animal and passed to another animal by sting or bite.

vertebrate: An animal that has a backbone.

FOR MORE INFORMATION

BOOKS

Amstutz, Lisa J. *Mammals*. Oxford, UK: Raintree, 2018.

Daniels, Patricia. *Mammals*. Washington, D.C.: National Geographic Kids, 2019.

WEBSITES

DK Find Out!

You can learn all about different kinds of mammals here. There are also some games to explore!

www.dkfindout.com/us/animals-and-nature/mammals/

World Wildlife Foundation

Is there an animal you want to learn more about? You can search for it at this website!

www.worldwildlife.org/species

Publisher's note to educators and parents: Our editors have carefully reviewed these websites to ensure that they are suitable for students. Many websites change frequently, however, and we cannot guarantee that a site's future contents will continue to meet our high standards of quality and educational value. Be advised that students should be closely supervised whenever they access the internet.

INDEX